# DRAGONFLIES UP CLOSE

AMANDA VINK

New York

Published in 2020 by The Rosen Publishing Group, Inc.
29 East 21st Street, New York, NY 10010



First Edition

Editor: Elizabeth Krajnik
Book Design: Michael Flynn

Photo Credits: Cover, p. 1 Digital Images Studio/Shutterstock.com; (series background) Karuka/Shutterstock.com; p. 5 Rudmer Zwerver/Shutterstock.com; p. 7 Film photo/Shutterstock.com; p. 9 piotreknik/Shutterstock.com; p. 11 Bonnie Taylor Barry/Shutterstock.com; p. 13 yanikap/Shutterstock.com; p. 15 LindaSh/Shutterstock.com; p. 17 Klimek Pavol/Shutterstock.com; p. 19 AlekseyKarpenko/Shutterstock.com; p. 21 R0macho/Shutterstock.com; p. 22 Butterfly Hunter/Shutterstock.com.

Cataloging-in-Publication Data

Names: Vink, Amanda.
Title: Dragonflies up close / Amanda Vink.
Description: New York : PowerKids Press, 2020. | Series: Bugs up close! | Includes glossary and index.
Identifiers: ISBN 9781725307865 (pbk.) | ISBN 9781725307889 (library bound) | ISBN 9781725307872 (6 pack)
Subjects: LCSH: Dragonflies–Juvenile literature.
Classification: LCC QL520.V56 2020 | DDC 595.7'33–dc23

Manufactured in the United States of America

CPSIA Compliance Information: Batch #CWPK20. For Further Information contact Rosen Publishing, New York, New York at 1-800-237-9932.

# CONTENTS

## Dragonfly Basics

Dragonflies have lived on Earth for around 300 million years. Today's dragonflies look almost exactly the same as the first dragonflies did. There are about 3,000 species, or kinds, of dragonflies. However, all species of dragonflies have things in common.

# From Egg to Adult

Dragonflies may lay their eggs on plants or in water. When an egg **hatches**, the wormlike creature that comes out is called a larva. Over time, the larva will **molt** until it becomes an adult. Unlike other **insects**, dragonflies don't have a stage between larva and adult.

## Parts of a Dragonfly

A dragonfly's head has antennae, eyes, and mandibles, or mouthparts. Dragonflies have compound eyes, which means their eyes are made up of hundreds of lenses that make one picture in the **brain**. They also have ocelli, which are a type of simple eye that senses movement.

antennae
eye
mandibles

## The Thorax

The dragonfly's middle part is called the thorax. This is where the dragonfly's two pairs of wings and three pairs of legs are. At the end of each leg is a claw. The thorax may have three parts: top, shoulders, and sides.

wing
thorax
claw
leg

## Long Wings

You can tell what species a dragonfly is by the **veins** in its wings. Dragonfly wings have a part called the stigma, which is found near the tip of each wing. Their wings also have a part called the nodus, which is found in the middle of each wing.

stigma
nodus

## The Abdomen

The dragonfly's end part is called the abdomen. The abdomen has 10 parts that are all connected. This allows the abdomen to bend up and down. The dragonfly's **reproductive** parts are in the abdomen. Males and females have different reproductive parts.

abdomen

## Dragonfly Diet

In their larval stage, dragonflies eat many things, including fish, bugs called mosquitoes, insect larvae, and other dragonfly larvae. As adults, dragonflies eat insects. They help control the number of mosquitoes. One adult dragonfly may eat hundreds of mosquitoes each day.

## On the Food Chain

Dragonflies play an important part in the food chain. They eat and help control the number of other insects, including mosquitoes. Ducks, toads, and fish eat dragonflies in their larval stage. Bees, birds, fish, frogs, and some meat-eating plants eat adult dragonflies.

## Dragonfly Habitats

Dragonflies live in many different **habitats**. Young dragonflies most often live in slow-moving freshwater habitats such as streams and ponds. Adult dragonflies like to live near freshwater too, but they fly farther away to find food and to **migrate**.

## Beautiful Bugs

Dragonflies move so fast it's hard to see them. When they rest, though, it's easy to notice their beauty. In some places, people create freshwater ponds so dragonflies will visit. Whether they're warming themselves in the sun or zooming through the air, dragonflies are a sight to see!

# GLOSSARY

**brain:** An organ in the head that controls movement, thoughts, feelings, and more.

**habitat:** The natural home for plants, animals, and other living things.

**hatch:** To break open or come out of.

**insect:** A small animal that has six legs and a body formed of three parts and that may have wings.

**migrate:** To move from one area to another for feeding or having babies.

**molt:** To lose a covering (of skin, hair, etc.) and replace it with a new one.

**reproductive:** Relating to reproduction, or the act or process of making babies.

**vein:** A tube in the body that carries blood back to the heart.

# INDEX



# WEBSITES

Due to the changing nature of Internet links, PowerKids Press has developed an online list of websites related to the subject of this book. This site is updated regularly. Please use this link to access the list: www.powerkidslinks.com/buc/dragonflies